Math Concept Reader

Cou
in the City

by Jean Sharp
Illustrations by Lorin Walter

Developed for Harcourt, Inc., by Gareth Stevens, Inc.

Printed in the United States of America

ISBN 13: 978-0-15-360275-7
ISBN 10: 0-15-360275-9

1 2 3 4 5 6 7 8 9 10 039 16 15 14 13 12 11 10 09 08 07

Harcourt
SCHOOL PUBLISHERS

We go to the city.

We count many things.

I count 9 buildings.

I count 7 streetlights.

I count 5 yellow taxis.

I count 2 city buses.

We see 3 firefighters.

We see 4 police officers.

There are 10 children.

There are 6 teachers.

We count 17 balloons.

We count 13 dogs.
There are 0 cats.

We count 16 boats.

We count 20 ducks.

I count 12 muffins.

I count 11 books.

I count 8 green apples.

I count 14 red flowers.

It is night. I count 30 lights.

I count 24 stars.

There is 1 train.

We ride it home.

Glossary

bakery 

city 

firefighter 

fruit stand 

muffin 

police officer 

taxi